Frederick Henry Duryea Crane

A Contribution to the Knowledge of Tellurium

Frederick Henry Duryea Crane

A Contribution to the Knowledge of Tellurium

ISBN/EAN: 9783337222215

Printed in Europe, USA, Canada, Australia, Japan

Cover: Foto ©Thomas Meinert / pixelio.de

More available books at **www.hansebooks.com**

A CONTRIBUTION TO THE KNOWLEDGE OF

TELLURIUM

A DISSERTATION

PRESENTED TO THE BOARD OF UNIVERSITY STUDIES

OF THE

JOHNS HOPKINS UNIVERSITY

FOR THE DEGREE OF

DOCTOR OF PHILOSOPHY

BY

FREDERICK HENRY DURYEA CRANE

1898

Contents

Acknowedgement	
Introduction	1
The raw material	1
Extraction of tellurium	2
Precipitation of tellurium	3
Precipitation by magnesium	5
Detection of tellurium	6
Detection of selenium	7
Precipitation of tellurium by ferrous sulphate	7
Purification of tellurium	10
Separation of selenium by tellurium	15
Further purification of tellurium	18
Determination of tellurium	19
The method and results	24
A form of the dioxid	28
Decompositions of the tetrachlorid	29
Notes	30
Conclusions	31
Biographical sketch	

This work was begun and carried on under the direction of Dr. Ira Remsen, and I thank him for his teaching and counsels. I am also sincerely grateful to Dr. Harmon N. Morse and Dr. Edward Renouf for their instructions; and to Dr. William H. Howell and Dr. Joseph S. Ames for instruction in my subordinate subjects. I wish also to express my obligations to Dr. Louis Duncan and Dr. Hermann S. Hering for their kind assistance.

In spite of its one hundred years of existence as a known substance, the nature of Tellurium is almost as obscure as in the time of Berzelius. This has been largely due to its scarcity. It could only be gotten in small quantity by length processes of extraction applied to large masses of crude ores. And, when obtained, it was still uncertain that it was the same substance that had before been described as derived from other sources, since the methods of purification were various, and imperfect.

Moreover, the results of determinations of its atomic weight were conflicting.

The rise of the electric refining industry has placed within reach the dross which is produced in large quantities in the final purification of the precious metals, and it is with tellurium from this source that the following work has been conducted.

The raw material. Large quantities of low grade gold and silver ores found in our Western States are now profitably worked by smelting them with copper ores; the matte thus obtained being again smelted, rolled into plates, and the copper electrically refined.

In the tanks in which this is done, the precious metals and the impurities collect as a black slime (1).

(1) C. Whitehead, Jour. Am. Chem. Soc. 17, 849.

This is smelted with sand and niter and the silver and gold reduced and recovered. The slag is lixiviated, and the solution acidified with sulphuric acid. The voluminous white precipitate is allowed to settle; the acid liquid is then filtered off, and the mud dried at a low heat. A chalky, friable mass, with some harder portions due to greater heat in drying, results. It contains much silica, tellurium mostly as tellurite, with a little tellurate due to the drying, and some selenium, antimony, arsenic, copper and other metals. There are varying amounts of iron and aluminium from the flux. Potassium is present in large quantity, but not sodium.

Extraction of the tellurium. From this white substance the tellurium is readily extracted by repeated leaching of the powdered material with strong commercial hydrochloric acid. Heating the acid does not appear to be of much advantage. The best results are obtained when large quantities of the acid are used.

In order to filter the slimy mud at all rapidly, a
good pump and as large a filtering surface as possible
should be used. To make such a surface a 3.5 cm. Witte
plate is put in a large funnel and covered with a rather
thick layer of glass beads. On this is placed a layer of
asbestos in the usual manner. This device allows lateral
passage of the filtrate, and gives a much larger effective
surface than the combined area of the holes of a plate alone.

The bright yellow hydrochloric acid solution which
is obtained is evaporated to a convenient consistency,
depending on its intended use. It contains considerable
quantities of the metals which occur as impurities (2),
(2) Keller, Jour. Am. Chem. Soc. 19, 778.
as well as the tellurium and selenium.

Precipitation of tellurium. Since the work of
Berzelius the use of sulphur dioxid as a precipitant of
tellurium has ordinarily been advised, although it has
been known equally long that the action was never quite
complete and that traces of other metals, as well as all
the selenium, came down also. Its exceeding convenience
however was a strong reason for its use in the present

instance for at least the preliminary precipitation. It was found that, in all probability, the main reason for the incompleteness of its action was the very rapid increase in the ratio of acid to unprecipitated tellurium in the solution; two-thirds of this being due to the hydrochloric acid set free, and one-third to the sulphuric acid formed.

If these could be removed the precipitation should go on. Evaporation sufficed for the hydrochloric acid, and an additional quantity of tellurium was obtained, but the increase of the sulphuric acid soon stopped the reaction.

The practical removal of the acids by neutralisation was then tried, and it was found that the addition of an alkali or alkali carbonate resulted in a renewed precipitation. At this point Mr. R.L. Whitehead advised the use of acid sodium, (or potassium), sulphite; and his suggestion was thenceforward followed.

But even with this very efficient reagent the action in the cold is never quite complete. It was found that if a solution in which neither more acid sulphite, or more hydrochloric acid added to decompose some of the excess of acid sulphite already present, will produce a further

precipitate, be heated to boiling, that there is then a further precipitation without any addition of reagents.

It is better to remove the first precipitate before boiling, as the action is then more evident. This new precipitate gives the qualitative reactions for tellurium, and differs from the first precipitate only slightly in tint, its formation is probably due to the decomposition of some alkali tellurate at first formed through mass action. By this means, then, the tellurium was obtained, mixed with selenium, and a little of the other impurities.

Precipitation of tellurium by magnesium. In devising a method for the more accurate estimation of tellurium it was found that metallic magnesium would completely precipitate tellurium from a solution of the tetrachlorid in hydrochloric acid. The excess of hydrochloric acid should be as small as possible, and a slight excess of magnesium should be added. This latter may be destroyed by vigorous boiling, when it decomposes the water; the magnesium hydroxid is then removed by acidifying very slightly with acetic acid. The well washed tellurium precipitated by this method seems to be less easily oxidised.

Detection of tellurium. The apparent completeness of the action of acid sodium sulphite led to the idea that it would detect small quantities of tellurium and do this more rapidly and efficiently than former methods. To test this, a weighed portion, (0.107 gm.), of tellurium was dissolved in hydrochloric acid by the aid of as little chlorine gas as possible, and the solution gently heated to expel any excess of chlorine. It was then diluted with a little hydrochloric acid and then with water to 500 cc. Ten cc. of this solution gave a marked precipitate, so 10 cc. were diluted to 110 cc., and the same quantity of this also gave a strong reaction, the dilution was repeated in a similar manner. An effective surface of about one-half a square centimeter of white filter paper was used to collect the precipitate, and on this the layer of black tellurium was plainly visible; the amount present per cc. of solution being 0.00000214 gm., and the total quantity used being 0.0000214 gm.

And the presence of tellurium was still discernable when the diluted solution contained only 0.000000214 gm. per cc. and there was present but 0.00000214 gm.

It appears probable that this limit is due simply to
the fact that we have here nearly reached the physiological
limit of seeing black on white.

Detection of selenium. Mr. Edward Keller has
recently called attention to the precipitation of selenium
by ferrous sulphate,(3), and this reaction was tested in

(3) Jour.Am.Chem.Soc.19,771.

like manner; 0.1139 gm. of selenium being weighed out,
dissolved and diluted. The precipitated selenium colored
the paper with the bright, characteristic red when the
dilution was 0.00000207 gm. per cc., ten times that quantity
being used. And at a dilution of 0.000000207 gm. per cc. the
precipitate from 10 cc. was more easily seen than that of
tellurium at about that dilution.

Furthermore, the precipitation of selenium at a
dilution of 0.00000207 gm. per cc. is independent of a
little more than one hundred times as much tellurium as
tetrachlorid.

Precipitation of tellurium by ferrous sulphate.
It is stated by Keller (4) that ferrous sulphate does

(4) Keller, loc.cit.

not precipitate tellurium. The statement is evidently made
with respect to a solution of the tetrachlorid in hydrochloric acid, and if there is certainly only the tetrachlorid
present, it is quite true.

errous sulphate, however, precipitates tellurium from
a solution of the tetrachlorid in hydrochloric acid if
it has been boiled vigorously for some time, the evaporating
acid being replenished; or if it is boiled, heated, or
remains for some time in contact, with tellurium.

This seems to be due to the presence of tellurium
dichlorid. But tellurium dichlorid, according to Rose (5)

(5) Rose, Pogg. Ann. 21, 443.

breaks down in the presence of acids into tellurous acid
and tellurium. To test this the two chlorids of tellurium
were made, first by the direct action of chlorine on
tellurium, and then by the action of this tetrachlorid on
the proper amount of tellurium. The solution which was obtained by treating the dichlorid in the cold with hydrochloric acid gave a black precipitate with ferrous sulphate
to a slight extent, and a solution made with the hot acid
gave it much more markedly.

A mixture of the two chlorids treated with the acid gave a very decided precipitate, and a solution of the dichlorid in a strong solution of the tetrachlorid in hydrochloric acid, gave it most strongly of all. In all these experiments the known decomposition of the dichlorid into free tellurium was prominent and evidently nearly in theoretical quantity, so all the solutions were well filtered before the ferrous sulphate was added.

The actual mass of the precipitated tellurium is small in comparison with the quantity present as chlorid, but the deep black color and the very fine state of division make the appearance a marked one. The black precipitate collects and settles after a time, and seems to diminish in bulk.

This is no doubt the same sort of change as that which occurs in tellurium precipitated by sulphurous acid or acid sulphite, and that is likely akin to the changes in precipitated selenium and tellurium when heated.

Furthermore, a boiling solution of tetrachlorid in hydrochloric acid will dissolve a little tellurium. In order to have the action complete, there must be but little of it, and that in a state of fine division.

This action may account in part for the apparent
diminution of the precipitate just mentioned.

It is hard to keep a concentrated solution of the
chlorid in a condition in which it gives no precipitate
with ferrous sulphate, but it may be quickly brought to
such a condition by passing through it a few bubbles of
chlorine, the excess of which is soon lost by allowing the
solution to stand open a short time.

Purification of tellurium. The purification
of tellurium in quantity appeared to be a question of
getting rid of selenium on the one hand, and of various
more metallic elements on the other. And it seemed advisable to do this as simply as possible in view of the
uncertainty as to the individuality of the substance.

Suspicion has been cast, at various times, on all methods
which require repeated evaporations, distillations or
crystallisations, or in which nitric acid must be removed
by prolonged heating. Further, it has been directly alleged
that distillation in hydrogen gave a product with a
lower ~~relative~~ atomic weight (6).

(6) rauner, Jour. Chem. Soc. 55, 392.

The method of Keller works well for selenium, if
precaution is taken to prevent a reversal of the action
through the mass action of the tellurium tetrachlorid on
the precipitated selenium, which will be refered to below.

But it has the great disadvantage of burdening the
solution with iron salts and sulphuric acid, which latter
always makes the precipitation of tellurium more difficult.

The method of Stolba (7) which depends on the reduction
(7) Stolba, Jahresbericht 1873, 214.
of a tellurite in alkaline solution by boiling with
glucose; which reduction is claimed to begin and be completed before a similar action with any selenite present,
appears to work as described. But in the presence of the
excess of glucose and its decomposition products it is
very hard to determine the limits of the reactions. Pure
tellurium is unquestionably gotten at first, but the
uniformity of the reaction remained in doubt. This process
could doubtless be made to work well, it was not used
since the process to be described seemed simpler.

As only a small fraction of the more metallic elements
is carried down by the tellurium, it seemed probable that

several precipitations would remove practically all of these. Consequently it was ~~required to~~ repeatedly convert the precipitated tellurium into the tetrachlorid.

The direct action of gaseous chlorine on dry tellurium is very rapid, but much heat is evolved, enough, in fact, to volatilise a part of the chlorid unless the process is carried on quite slowly. The action of gaseous chlorine on tellurium suspended in strong hydrochloric acid is, on the other hand, quite slow; and rapidly becomes slower as the quantity of dissolved chlorid increases, and is very slow if the tellurium is at all compact.

It has long been known that tellurium is much more metallic than selenium, and that selenium is the first to be precipitated by sulphur dioxid. In fact this method has been suggested for their separation. Now it seemed probable that, if the mixed precipitate could be subjected to the action of nascent chlorine, by being made the positive pole of an electrolytic cell containing hydrochloric acid, ~~that~~ not only would the tellurium be rapidly converted into the chlorid, at a low temperature, but that also the more metallic tellurium would be first attacked, to the practical if not total exclusion of the selenium.

Preliminary trials on a small scale having shown the rapid action of the nascent chlorine, a larger apparatus was made. This consisted of a 750 cc. funnel with the stem ground square off. In the bottom of the funnel position was placed a button of commutator carbon about 2.5 cm. in diameter, to which was soldered a small brass nut.

The button was sloped to fit the funnel and jacketed with a doubled rubber tube. Through the stem of the funnel was passed a heavy copper wire, threaded at both ends, and having on the lower end a small nut. The upper end of this wire was screwed into the nut on the button, and the button then drawn down tight by the lower nut acting against the lower end of the funnel stem. This formed the positive pole, the current being taken in on the wire.

The negative pole was formed of a sheet of thin copper of some 30 sq.cm. which was soldered to a convenient wire.

This pole was never in the acid but a moment before the current was on, and was removed at once when it was off, so that it was not attacked by the acid. To prevent the deposition of tellurium it was inclosed in a little porous cell which was hung on the edge of the funnel.

A small automatic drip supplied pure hudrochloric acid just fast enough to keep the surface of the liquid in the cell about 0.5 cm. above that in the funnel, so that there was always a flow of pure acid through the cell away from the negative pole. It was expected that the sloping sides of the funnel would keep the unattacked tellurium always at the bottom, but as the specific gravity of the solution increased, the solid did not sink rapidly enough, so a glass rod, mechanically turned, was put in to act as a stirer. When somewhat heated from the passage of the current this arrangement had a very constant resistance of a little more than half an ohm, and, with the current employed, used about nine amperes.

Practically every trace of the chlorine liberated was absorbed, and consequently the speed of solution depended directly on the current. The first charge of the cell was not quite all dissolved when it became neccessary to stop the action over night, and the entire contents of the cell was removed to another vessel. At the same time a portion of the liquid was taken out and tested for selenium, which was found to be present. Evidently the

tellurium was not exclusively or primarily attacked, as some of it visibly remained. The next morning this test was repeated, and no selenium was found. And the reddish color of the sediment indicated that it had been precipitated. This electrolytic chlorine method of solution is very efficient and has the marked advantage of adding no extraneous substance. It is more rapid, of course, as the tellurium is the more finely divided.

    Separation of selenium by tellurium. Further experiments showed that if to a mixture of selenium dichlorid and tellurium tetrachlorid dissolved in hydrochloric acid tellurium is added; or a portion of the tellurium precipitated; and the mixture allowed to stand for some time, or better, and much more expeditiously, boiled; all the selenium will disappear from the solution, and may be found in the sediment, with the excess of tellurium. If the action takes place at the ordinary room temperature, the selenium can easily be seen on account of its color, but if heat is used the well known change to the black form will occur, and it can not be distinguished from the tellurium.

 It may be readily detected, however, by filtering off

the black residue and dissolving it in a little hydrochloric acid by the aid of chlorine, when the selenium may be again precipitated. If this reaction be carried on upon a microscope slide, it can be well seen and makes a rather pretty effect when magnified to about 80 diameters.

On a large scale it is preferable to heat the solution of mixed chlorids gently just below the boiling-point, and to have the tellurium in as fine state of division as possible and well mixed with the fluid. To attain both these ends it is advisable to precipitate some of the tellurium, either by leading in a little sulphur dioxid, or adding a little acid sulphite. The turbid liquid which results does not clear for some time by settling.

That this process is complete is shown by the fact that it not only removes selenium from tellurium tetrachlorid solutions so completely that none could be detected by the ferrous sulphate test, but also removed selenium completely from a solution of selenium dichlorid in hydrochloric acid, previously precipitated tellurium being employed. In this experiment tellurium is found in solution which shows conclusively a replacement or an exchange.

If, on the other hand, selenium be gently heated in a solution of tellurium tetrachlorid for some hours, the addition of ferrous sulphate to the filtered fluid will show that a very small part of the selenium has gone into solution. This minute amount may be entirely removed by the treatment with tellurium. This seems to explain the observation that in working with quite large quantities of the mixed chlorids in solution the action of ferrous sulphate even in obvious excess did not seem to be quite complete if a few hours elapsed before the precipitated selenium was filtered off. But the solution could not be made to take up selenium if there was a little tellurium present and it was kept well mixed up. The phenomenon appears to be, to some extent at least, one of mass action.

In using this process for the removal of selenium the precipitate, containing the excess tellurium is filtered off and redissolved, either by nascent chlorine, or by conducting into it, suspended in hydrochloric acid, a stream of that gas. The selenium is then easily removed by ferrous sulphate and the tellurium by acid sulphite. It is well to avoid the use of nitric acid.

The further purification of tellurium. The
logical sequence to boiling the mixed chlorids with
tellurium in order to precipitate selenium or any less
metallic elements was to boil the precipitated tellurium
with a reserved portion of that solution from which it
had been obtained, so that any more metallic elements
which might have been carried down would be dissolved, and
precipitate an equivalent quantity of tellurium. This
was done, although several precipitations in the ordinary
manner seemed to have already removed those traces of
metals which have been previously noted as occurring in
the crude material.

The tellurium prepared in this manner has the usual
appearance and reactions, and could be completely distilled
in a current of hydrogen, leaving only a slight residue
of carbonaceous material probably derived from the filter
papers. It was noted that there was very little tendency
to form hydrogen telluride with pure, dry hydrogen.

This process should tend to separate the suspected
homologue of tellurium of greater relative weight, and
although no definite indications of such a substance have

been met with up to the present, work will be continued along this line if time and oppertunity are found.

Determination of tellurium. There has never been any satisfactory method of determining tellurium. Most observers have either determined other elements in the tellurium compounds, or precipitated the tellurium, treated it with nitric acid and converted the compound thus formed into the dioxid by heat. This has been the most exact method.

But Brauner (8) notes the brown decomposition which

(8) Brauner, loc. cit.

Staudenmaier (9) had observed, and further claims that the

(9) Staudenmaier, Ztschr. anorg. Chem. 10, 206.

nitric acid is not wholly driven off before a part of the dioxid is volatilised. A method of avoiding the brown color has lately been given by Norris and Fay (10), but

(10) Norris and Fay, Am. Chem. Jour. 20, 278.

nothing is said as to the removal of the nitric acid, and there is visible volatilisation. But gravimetric determinations by the method about to be described showed that the dioxid prepared by that method was quite pure. However, it is doubtful if nitric acid can be removed without some volatilisation.

o volumetric method which does not require a correction term,with the exception of that of Norris and Fay has been devised. The objections to the direct precipitation and weighing of the tellurium were alleged to be two;that it was not possible to precipitate all the tellurium from an acid solution,and that after precipitation it was oxidised so much in drying that the results were variable (11).

(11) Brauner,loc.cit.;Norris and Fay,loc.cit.

In view of the extreme delicacy of the acid sodium sulphite when used to detect tellurium,it seemed advisable to try it as a precipitating agent in quantitative work.

There is no doubt that free acid does prevent the complete action of sulphur dioxid,but there is no evidence that neutral alkali salts in solution do this,if the solution is heated;at least qualitative experiments failed to show such an action.

So it is only necessary to add enough acid sulphite to give up soda to the acids present and set free. A slight excess of sulphite does no harm,and an excess of sulphur dioxid is readily removed.

After the acid sulphite ha been added, it is advisable to let the mixture stand for a time, usually, for convenience, over night, as the precipitate seems to be in better mechanical form when this is done. But the action may be completed at once by heating. At any rate, the liquid with the precipitate still in it should be raised to the boiling-point for a short time.

This is to gain two ends; to insure the total precipitation of the tellurium, which is never quite complete in the cold when large amounts are being handled, (as has been noted); and to cause some definite but undetermined change in the precipitate. As a result of this it becomes much easier to filter and wash, and less liable to oxidation.

This change, as has been said, seems to be like the long-known change in selenium. It is hard to describe, but easily recognised, although it consists only of the merest change in tint, and a difference in the manner in which the precipitate settles.

The precipitate is then collected on a Gooch filter and very thoroughly washed, care being taken that it is always covered with water. If speed is more requisite

than exactness it may now be pumped as dry as possible,
quickly dried in an air-bath and weighed. It will oxidise
to some extent, but by no means seriously.

The oxidation can easily be detected and the extent
of it quite accurately judged by removing the precipitate
with the filter surface, after weighing, and putting it in
a beaker on a white surface, or in an evaporating-dish,
and moistening it with a few drops of strong hydrochloric
acid. A yellow tint of dissolved chlorid appears at once
only if oxidation has occurred, as unoxidised tellurium
does not color hydrochloric acid. But if this preparation
be allowed to stand, a color will appear, as the air acts
readily on the acid mud. To determine how far this color
could be used as a criterion of the presence of oxid and
its amount, a weighed portion, 0.0068 gm., of pure tellurium
tetrachlorid was dissolved in strong hydrochloric acid
and diluting. The yellow color of a layer 0.25 cm. thick was
plainly visible when 0.00013 gm. of tetrachlorid was present
in each cc.

To see if the method without special precautions in
drying was even nearly correct, portions of a pure precipitated

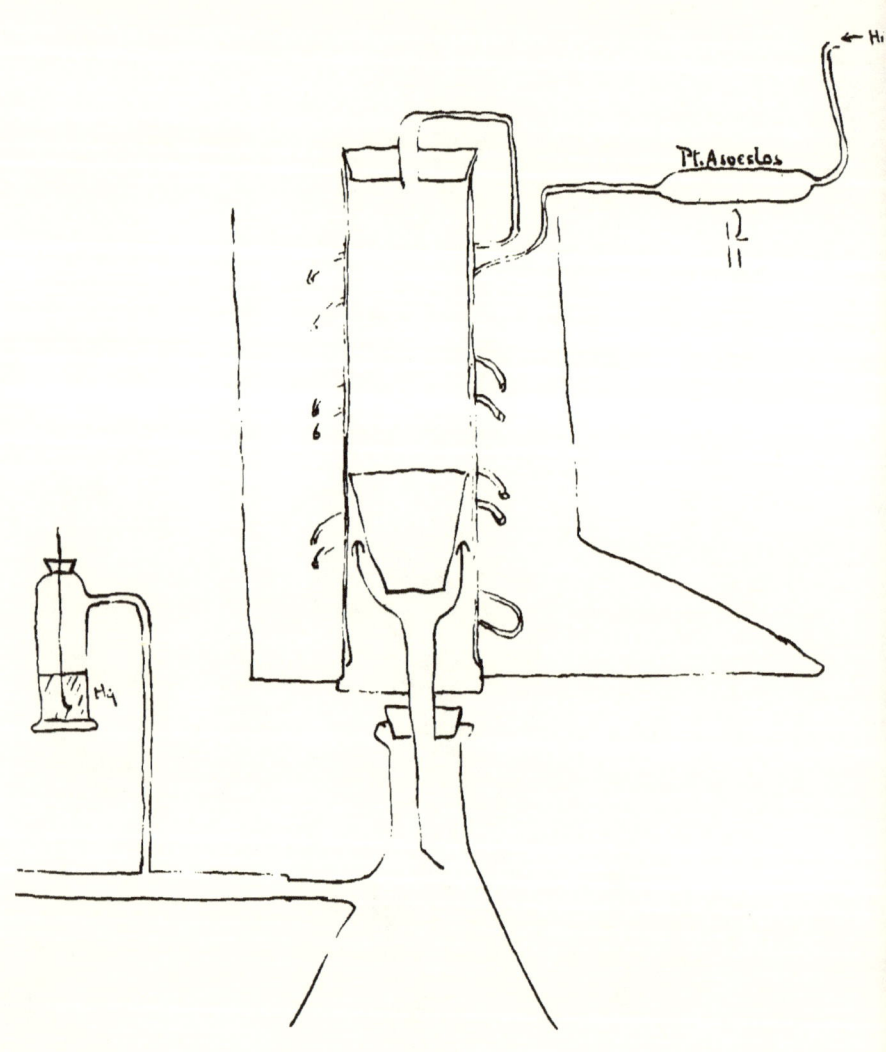

tellurium in which very little oxidation had occurred
were weighed out, dissolved in a little hydrochloric acid
by the aid of a little chlorine gas and treated as stated
above.

Taken	Found	Gain	% Recovered
0.1768 gm.	0.1779	0.0011	100.6
0.1285	0.1289	0.0004	100.3

Probably practice in its use would make this method
exact enough for all practical purposes. From the article
by Mr. Keller already referred to it appears that no
special precautions in drying are used at the works of the
Baltimore Electric Refining Company.

To prevent, if possible, any oxidation, the filtering
crucible was enclosed in a cylinder through which dry,
pure hydrogen which had just been passed over hot platin-
ised asbestos was slowly drawn under slightly diminished
pressure. The device was so arranged that it was not
neccessary to remove it from the filter flask, and a
T-tube in the pump connection was joined to a mercury
saftey valve by which both the pressure and flow were
regulated.

The cylinder and its contents were heated by jacketing them with a strong solution of chlorid of calcium, after the manner of a hot water funnel, and this bath was also used to warm the entering stream of hydrogen.

This arrangement prevents oxidation as far as can be detected by the coloration of hydrochloric acid, and as the results seem quite constant, it probably does not occur to a measurable extent. There is no evidence to show that tellurium which has been boiled with water can decompose water in the presence of hydrogen, so that the essential thing is to have pure hydrogen.

The method. The entire process is as follows. The given compound is so treated that the tellurium is in the form of the tetrachlorid, avoiding if possible, the use of nitric acid, and as little hydrochloric acid as possible being used. The solution is now diluted with water, but not to such an extent that a white substance appears, although a little of this seems to do no harm.

The objection to a larger quantity is mechanical, the interior of the flocculent masses seems sometimes to be unattacked by the sulphurous acid.

A solution of acid sodium or potassium sulphite is now added. It should be moderately concentrated, and the quantity should be such as to neutralise nearly exactly the acids present and set free by the base of the reagent. The excess of sulphur dioxid will escape harmlessly, but anything more than a slight excess of the acid sulphite is detrimental. The mixture, which has turned dark, has been gently agitated during these manipulations. It is now diluted to about 50 to 75 cc. and allowed to stand for a time as the precipitate thus seems to form more evenly. But it may be warmed at once.

If too much sulphite has been inadvertantly been added the liquid above the precipitate should be decanted and reserved, and the precipitate washed once or twice and the decanted wash-water added to the other, and all this passed through the weighed filter later. The reason for this is that it is very hard to decant without getting some of the precipitate over, although it may remain invisible until it blackens the filter. If time is not a consideration, this decantation process may well be employed in all cases.

The precipitate, in about 75 cc. of water is now gently warmed, being stirred with a glass rod tipped with a bit of rubber tube, both to prevent "bumping" and to detach those portions which at times form a very marked "mirror" on the sides of the vessel; it is allowed to boil rapidly for a moment or two, and then set aside. A lit'le water from a wash-bottle may be added, as it aids the settling. The tellurium, which has changed in appearance, should settle very quickly. If it does not, it may be boiled a little more, but it is to be noted that differences in the concentration of the solution in which the precipitation took place may markedly influence the character of the precipitate.

The amount of material taken should be such that in the filter the tellurium will form a rather thin layer, even while still wet. The precipitate is now filtered through a weighed filter, care being taken to keep it always covered with water; and well washed, and since there is no reason to suspect any solvent action of the water, the washing should be thorough.

After the washing is completed the hydrogen supply
is connected at once, the last water being followed through
the filter by the hydrogen. But no effort need be made to
displace the small amount of air in the cylinder.

The temperature of the calcium-chlorid bath is now
quickly raised to about 110° and kept there till the
tellurium is dry. This must be judged from its appearance.

The filter is then removed to a desiccator, and, when cool,
weighed. There is no tendency to oxidise rapidly in the
air if thoroughly dry. The main danger to be avoided is
the accidental access of air to the wet tellurium.

Some of the results obtained by this method were as
follows,---

Tellurium dioxid from the nitrate, Te,127.6; 0,16.

Taken	Found	% Required	% Found.
0.1047 gm.	0.0838	79.95	80.03
0.1130	0.0904	79.95	80.00

Tellurium tetrachlorid distilled in carbon dioxid, Cl, 35.5.

Taken	Found	% Required	% Found
0.3379 gm.	0.1857	47.33	47.37
0.5678	0.2717	47.33	47.85

Tellurium oxychlorid, fused,

Taken	Found	% Required	% Found
0.62075 gm.	0.3689	59.46	59.43

Tellurous acid, from the chlorid,

Taken	Found	% Required	% Found
0.11985	0.0862	71.34	71.92

A yellow form of the dioxid. If a solution of tellurium tetrachlorid in hydrochloric acid which has stood for some time after the selenium has been precipitated by ferrous sulphate and removed; or a similar solution to which ferrous or ferric chlorid has been added is diluted with water, a white precipitate forms. If this is now allowed to stand for a time, the excess of water being decanted and replaced till it is not colored by the iron, a heavy yellow precipitate will appear. Or if the solution is poured directly into boiling water, the whole mass of precipitate will be colored yellow. The difference in specific gravity is so great that the remaining white substance may be almost completely removed by washing.

This substance is easily soluble in hydrochloric acid, and more difficultly in nitric and sulphuric acids.

It is readily and completely dissolved by alkalies to a colorless solution without residue. Under the microscope it is seen to consist of more or less regular yellow crystals, of the isometric system, octahedrons, sometimes modified by cubic faces. Among these were traces of a white substance which seemed to cling to the crystals.

It appears to be a form of tellurium dioxid, analyses giving the following percentages of tellurium, 79.46, 79.52, 79.51, 79.58, 79.46; calculated for TeO_2, 79.95.

The discrepancy may be due to the adherent white substance, or to a little iron, of which a trace may be found by the ferrocyanide and sulphocyanide tests. To this also may be due the yellow color, which is very persistent, not yielding to nitric acid till the substance is all in solution, but the solution is colorless, and deposits white crystals.

Decomposition of the tetrachlorid. If water is added to a portion of the distilled tetrachlorid of tellurium considerable heat is evolved and it passes into solution as a yellow liquid. if more water be added there is a curdy white substance formed, which settles and

after a time forms the crystalline dioxid. The white substance is tellurous acid, but the wash water from it constantly contains a trace of chlorine, and there may be a little of some sort of oxychlorid present.

But if there is also present another metal, as iron, and particularly if antimony or arsenic is present, there is a marked change in the character of the white substance, it remains unchanged in water for days, and may be washed, dried, and redissolved and reprecipitated several times without forming the dioxid. And no fixed amount of tellurium was found in various samples. It appears to be a double oxychlorid, and will be the subject of further investigation if possible.

Failure to form analogues of thiosulphate. The attempt was made to form substances analogous to thiosulphates by boiling sodium tellurite with sulphur, selenium and tellurium, but after several hours no traces of any action could be found.

Instability of the chlorids. Many minor observations, as well as those already noted, tend to show that the chlorids of tellurium tend to pass to a slight

extent from one to the other. This is especially true of
the tetrachlorid, which darkens on distilling in dry carbon
dioxid. It is to this that the high per cent, of tellurium
noted above is probably due. And even that distilled in
chlorin will give a slight cloud with ferrous sulphate
if it has been kept for a time.

Conclusions. Tellurium may be easily obtained
from certain wastes of electric refineries by extracting
with hydrochloric acid and precipitating with acid sodium
sulphite. Magnesium will also precipitate tellurium, and
has some advantages for qualitative work and also for
quantitative. Tellurium may be detected in very dilute
solutions by acid sulphite, and ferrous sulphate is equally sensitive with selenium, and is independent of tellurium
tetrachlorid. But ferrous sulphate precipitates tellurium
under some circumstances, which is probably because some
tellurium dichlorid is present in solution.

In order to avoid certain sources of suspected error,
and to add nothing but hydrochloric acid as reagent, a
method of dissolving tellurium by electrolytic chlorine
was devised, and was found to work rapidly and well.

While using this it was found that tellurium would remove selenium. The use of tellurium for this purpose rather than any other reagent is of advantage because it does not add any other elements to the solution, or permit the escape of any portion of the tellurium beyond control. This replacement method may be extended to remove more metallic elements.

Tellurium may be determined by precipitating with magnesium or acid sodium sulphite, boiling, drying in hydrogen and weighing as tellurium.

The decomposition of the tetrachlorid by water is not a simple matter, and becomes much more complex if other metals are present. The pure tetrachlorid gives tellurous acid which decomposes to tellurium dioxid.

The author was born in Boonton, New Jersey, October 15, 1872. He attended the public schools of Schenectady, New York, and entered Union College, from which he received the degrees of Bachelor of Arts, in 1893; and Master of Arts, in 1896. Since his graduation he has studied Chemistry, Physiology and Physics at the Johns Hopkins University.

www.ingramcontent.com/pod-product-compliance
Lightning Source LLC
Chambersburg PA
CBHW020232090426
42735CB00010B/1665